# Making Math Easy

## Column Addition & Subtraction Magic

ISBN: 978-1-960579-29-4



www.densingmethod.com

# Making Math Easy

## Column Addition & Subtraction Magic

Embark on an exciting journey through the world of column addition and subtraction with fun exercises and captivating illustrations. This workbook uses visual cues to enhance understanding seamlessly. It complements the principles in 'Hands-On Learning Adventures: Mathematics' from the Densing Teaching Method Book Series, making it an ideal companion when used with recommended poster boards.

In the accompanying book, you'll find step-by-step instructions for creating personalized math materials, fostering collaborative learning with your child. Large poster boards serve as effective tools for helping children memorize essential math facts related to addition and subtraction. This workbook provides essential practice in addition and subtraction, enriched with integrated visual cues.

With different approaches to arithmetic operations, you can discover the method that best suits your student. Children will learn that various methods can achieve the same results. Additional workbooks are available, focusing on their preferred method and offering extra practice opportunities. These resources are designed specifically to introduce written addition and subtraction, tailored for students who may find mathematics challenging.

# Column Addition & Subtraction Without Regrouping

This poster from the book **"Hands-On Learning Adventures: Mathematics"** symbolizes a tactile learning experience for understanding place value, column addition, and subtraction. The concept of place value can be challenging for children, especially when they transition to paper-based addition and subtraction, as they need to remember the correct order of operations.

The following math problems concentrate on basic addition and subtraction without regrouping, aiming to help children learn about place value and the order in which they need to add or subtract using the column method.

The cheetah symbolizes the ONES, being the fastest animal that goes down the tree first. The snail represents the TENS. Since it is slower, it goes down the tree second.

TENS
ONES
Always start with the ONES
3 3
+ 2 2
5 5

TENS
ONES
Always start with the ONES
3 3
+ 2 3

**Plus**

TENS
ONES
3 2
+ 2 3

TENS
ONES
2 3
+ 2 3

TENS
ONES
Always start with the ONES
3 6
+ 1 2

TENS
ONES
1 4
+ 5 3

## Plus

TENS
ONES
1 5
+ 3 2

TENS
ONES
4 4
+ 4 4

TENS
ONES
Always start with the ONES
6 6
+ 2 2

TENS
ONES
2 7
+ 2 1

**Plus**

TENS
ONES
7 5
+ 1 4

TENS
ONES
3 3
+ 6 5

TENS
ONES
Always start with the ONES
4 8
+ 2 1

TENS
ONES
6 7
+ 1 2

**Plus**

TENS
ONES
8 2
+ 4

TENS
ONES
5 3
+ 2 6

TENS
ONES
Always start with the ONES
4 0
+ 3 5

TENS
ONES
5 7
+ 2 1

**Plus**

TENS
ONES
3 5
+ 2 4

TENS
ONES
4 3
+ 3 4

TENS
ONES
Always start with the ONES
2 6
+ 6 2

TENS
ONES
1 7
+ 2 2

## Plus

TENS
ONES
5 4
+ 2 4

TENS
ONES
2 3
+ 3 6

TENS
ONES
Always start with the ONES
3 9
+ 5 0

TENS
ONES
2 5
+ 3 2

## Plus

TENS
ONES
6 5
+ 2 4

TENS
ONES
4 6
+ 3 1

TENS
ONES
Always start with the ONES
4 6
+ 4 0

TENS
ONES
3 5
+ 3 4

## Plus

TENS
ONES
6 3
+ 3 4

TENS
ONES
4 2
+ 4 1

TENS
ONES
Always start with the ONES
6 7
− 2 3
4 4

TENS
ONES
Always start with the ONES
6 7
− 2 4

## Minus

TENS
ONES
7 9
− 5 7

TENS
ONES
6 6
− 2 5

TENS
ONES
Always start with the ONES
6
5
4
3

TENS
ONES
6
9
4
3

## Minus

TENS
ONES
7
9
2
6

TENS
ONES
6
6
3
3

TENS
ONES
Always start with the ONES
5
7
2
2

TENS
ONES
7
7
2
2

# Minus

TENS
ONES
6
9
5
5

TENS
ONES
5
6
3
2

TENS
ONES
Always start with the ONES
9 9
1 8

TENS
ONES
8 5
3 4

## Minus

TENS
ONES
5 3
3 2

TENS
ONES
4 8
3 5

TENS
ONES
Always start with the ONES
9 7
7 6

TENS
ONES
6 5
3 2

# Minus

TENS
ONES
7 6
6 4

TENS
ONES
8 4
2 4

TENS
ONES
Always start with the ONES
5 7
3 1

TENS
ONES
6 0
1 0

## Minus

TENS
ONES
6 8
6 4

TENS
ONES
7 3
5 3

TENS
ONES
Always start with the ONES
5 6
− 2 1

TENS
ONES
6 3
− 1 1

## Minus

TENS
ONES
6 7
− 6 5

TENS
ONES
7 7
− 5 4

TENS
ONES
Always start with the ONES
5 4
− 2 3

TENS
ONES
6 2
− 2 1

## Minus

TENS
ONES
6 6
− 4 5

TENS
ONES
6 5
− 5 3

# Column Addition
# With Regrouping

**Introduction for Parents and Teachers:**

Invite your student to try an addition problem that involves regrouping. Remind them to begin with the ONES. If the sum of the ONES is 10 or greater, guide them to write the digit 1 (representing the TENS) inside the neck square and the other digit (representing the ONES) inside the yellow ONES square on the leg.

In our example, 24 + 38, we started by adding the ONES in the yellow squares (4 + 8 = 12). We placed the digit 1 of the TENS inside the neck and the digit 2 of the ONES inside the yellow square on the leg. Then, your child adds up all the numbers in the TENS place.

(In our example, 1 + 2 + 3 = 6). So, 24 + 38 = 62.

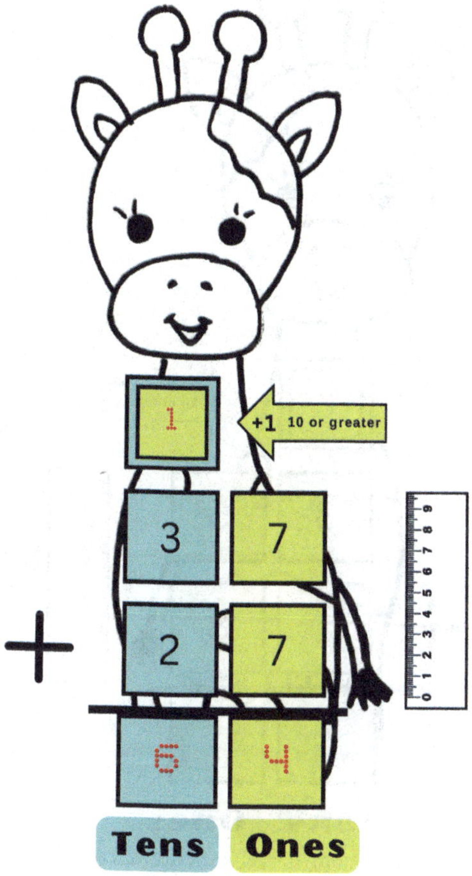
1
+1 10 or greater
3 7
+ 2 7
6 4
Tens
Ones

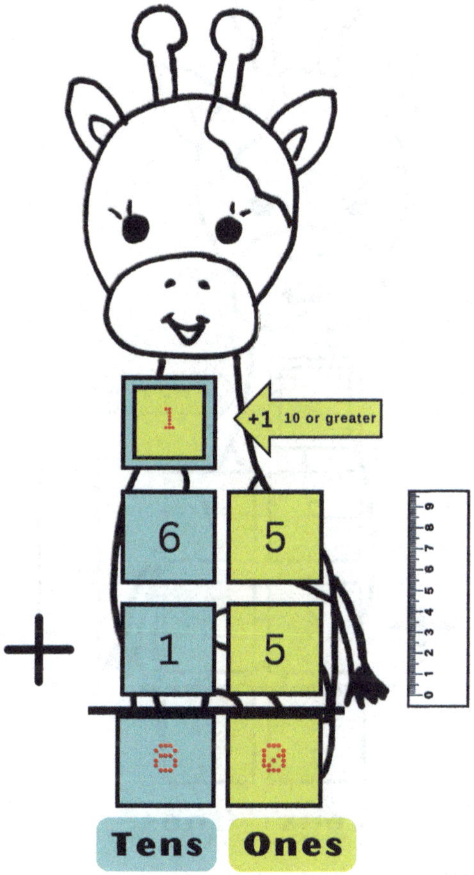
1
+1 10 or greater
6 5
+ 1 5
8 0
Tens
Ones

+1 10 or greater
1 6
+ 2 6
Tens
Ones

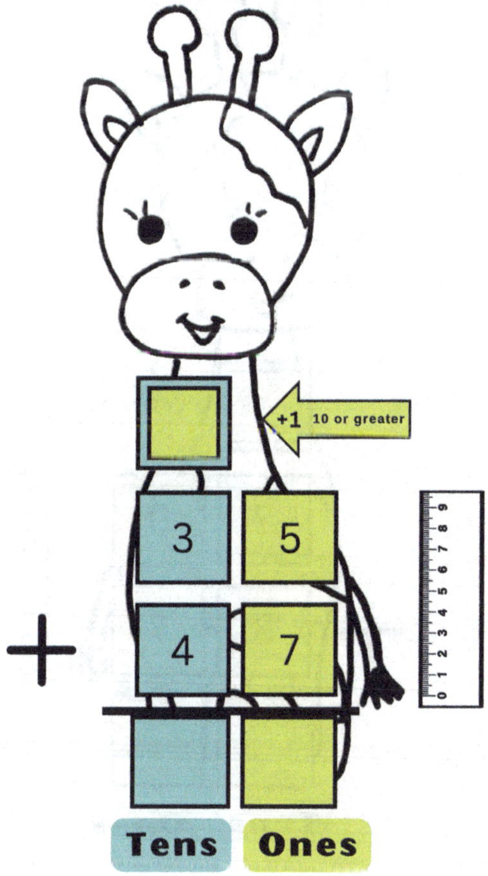
+1 10 or greater
3 5
+ 4 7
Tens
Ones

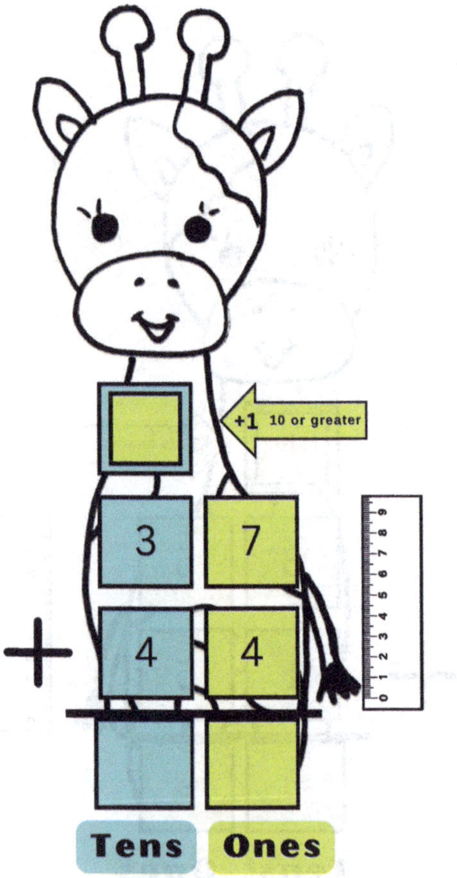
+1 10 or greater
3 7
+ 4 4
Tens
Ones
0 1 2 3 4 5 6 7 8 9

+1 10 or greater
3 5
+ 1 8
Tens
Ones
0 1 2 3 4 5 6 7 8 9

+1 10 or greater
2 9
+ 3 7
Tens
Ones
0 1 2 3 4 5 6 7 8 9

+1 10 or greater
4 3
+ 4 7
Tens
Ones
0 1 2 3 4 5 6 7 8 9

+1 10 or greater
1
6
+
2
5
0 1 2 3 4 5 6 7 8 9
Tens
Ones

+1 10 or greater
6
7
+
2
5
0 1 2 3 4 5 6 7 8 9
Tens
Ones

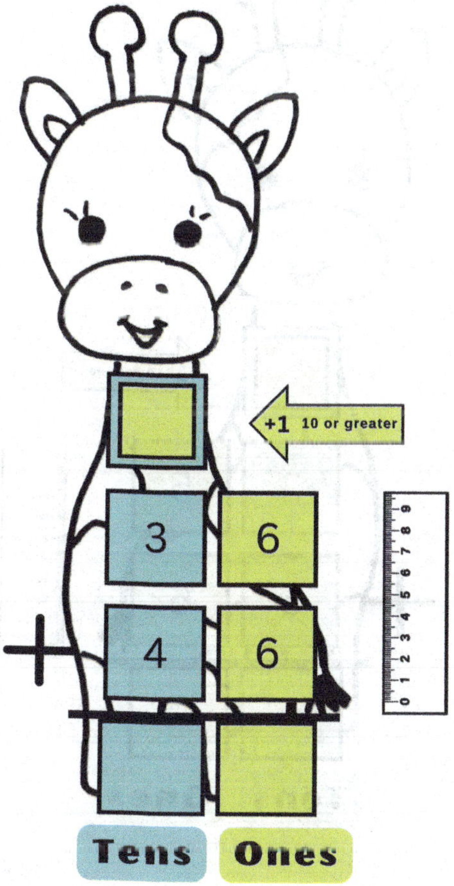
+1 10 or greater
3
6
+
4
6
0 1 2 3 4 5 6 7 8 9
Tens
Ones

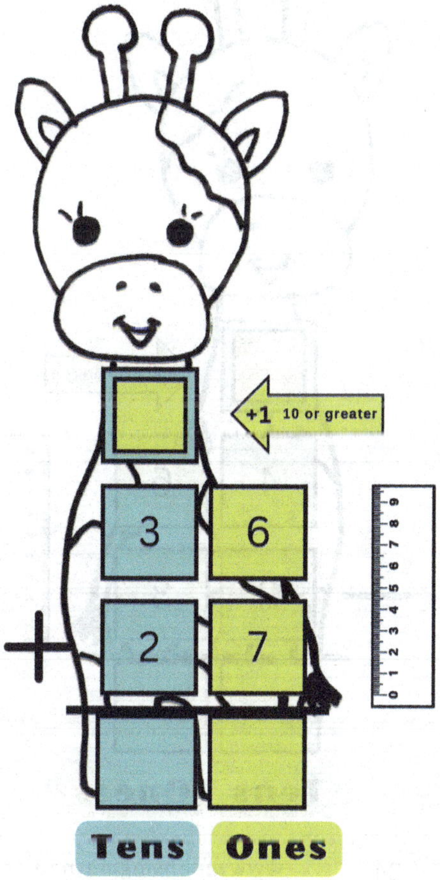
+1 10 or greater
3
6
+
2
7
0 1 2 3 4 5 6 7 8 9
Tens
Ones

+1 10 or greater
2
6
+
2
7
Tens
Ones

+1 10 or greater
6
6
+
2
5
Tens
Ones

+1 10 or greater
4
6
+
2
8
Tens
Ones

+1 10 or greater
3
5
+
5
6
Tens
Ones

+1 10 or greater
1 7
+ 2 6
Tens
Ones

+1 10 or greater
6 3
+ 1 9
Tens
Ones

+1 10 or greater
1 5
+ 2 7
Tens
Ones

+1 10 or greater
3 6
+ 4 8
Tens
Ones

+1 10 or greater
2
6
+
3
8
Tens
Ones

+1 10 or greater
3
5
+
2
5
Tens
Ones

+1 10 or greater
2
6
+
4
6
Tens
Ones

+1 10 or greater
1
5
+
1
7
Tens
Ones

+1 10 or greater
2 3
+ 3 9
0 1 2 3 4 5 6 7 8 9
Tens
Ones

+1 10 or greater
3 7
+ 2 8
0 1 2 3 4 5 6 7 8 9
Tens
Ones

+1 10 or greater
2 8
+ 4 7
0 1 2 3 4 5 6 7 8 9
Tens
Ones

+1 10 or greater
1 2
+ 1 8
0 1 2 3 4 5 6 7 8 9
Tens
Ones

+1 10 or greater
2
7
+
4
9
Tens
Ones

+1 10 or greater
3
4
+
3
8
Tens
Ones

+1 10 or greater
2
4
+
3
7
Tens
Ones

+1 10 or greater
1
3
+
3
8
Tens
Ones

# Flamingo Column Addition with Regrouping

**Introduction for Parents and Teachers:**

There are different ways to teach column addition. Find the method that works best with your child. These flamingo addition problems let students write the carried-over 1 below the numbers in the flamingo's leg. Remind students to always start with the ONES. If the sum of the ONES is 10 or more, they should write the 1 (for the tens) in the flamingo's yellow TENS square and the digit for the ONES in the yellow sum square on the bottom.

In the example 27 + 49, first add the ONES (7 + 9 = 16). Write the 1 from 16 in the yellow TENS square of the flamingo's leg and the 6 in the yellow sum square on the bottom.

Then, add the numbers in the TENS place (2 + 4 + 1 = 7). The final result is 76.

2 8
3 7
+ 1
+1 10 or greater
6 5
TENS
ONES

1 7
4 5
+
+1 10 or greater
TENS
ONES

2 3
2 9
+
+1 10 or greater
TENS
ONES

3 7
4 7
+
+1 10 or greater
TENS
ONES

2 5
5 6
+
+1 10 or greater
TENS
ONES

4 3
2 9
+
+1 10 or greater
TENS
ONES

7 4
1 9
+
+1 10 or greater
TENS
ONES

3 8
4 8
+
+1 10 or greater
TENS
ONES

1 8
5 3
+
+1 10 or greater
TENS
ONES

2 5
3 5
+
+1 10 or greater
TENS
ONES

2 8
4 7
+
+1 10 or greater
TENS
ONES

4 8
2 7
+
+1 10 or greater
TENS
ONES

5
7
1
4
+
+1 10 or greater
TENS
ONES

3
6
4
8
+
+1 10 or greater
TENS
ONES

3
5
1
8
+
+1 10 or greater
TENS
ONES

2
1
5
9
+
+1 10 or greater
TENS
ONES

3 7
5 5
+
+1 10 or greater
Tens
Ones

4 5
1 8
+
+1 10 or greater
Tens
Ones

1 3
4 9
+
+1 10 or greater
Tens
Ones

3 7
2 4
+
+1 10 or greater
Tens
Ones

1
6
7
9
+
+1
10 or greater
Tens
Ones

3
5
4
9
+
+1
10 or greater
Tens
Ones

6
8
1
8
+
+1
10 or greater
Tens
Ones

2
8
5
5
+
+1
10 or greater
Tens
Ones

# Column Subtraction With Regrouping

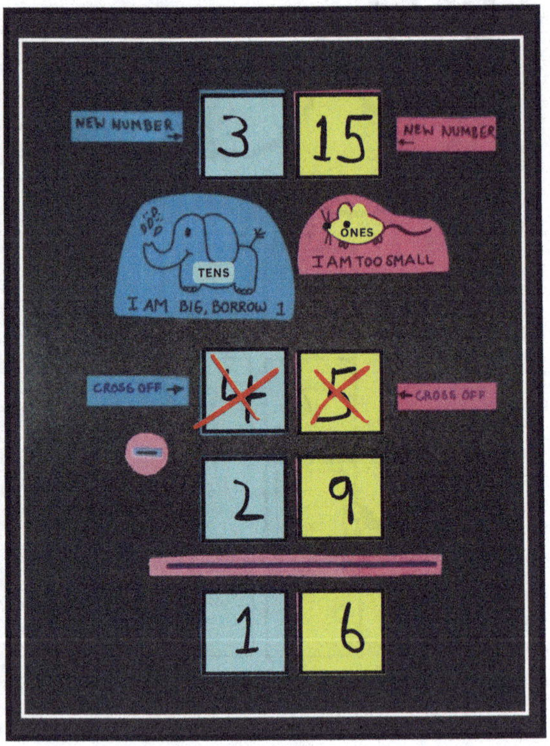


**Introduction for Parents and Teachers:**

**Minuend (first number 45) - Subtrahend (second number 29) = Difference (result 16)**

Invite your student to try a subtraction problem that involves regrouping. Remind your child to start with the ONES. In the column method of subtraction, we have to regroup the numbers if the digit of the minuend is smaller than the digit of the subtrahend. Children need to learn to borrow 1 from the next higher place and regroup the numbers. This colorful worksheet with an elephant and mouse theme from the book "Hands-On Learning Adventures: Mathematics" invites children to learn how to regroup by borrowing 1 TEN from the big elephant to the little mouse. In our example, we need to find the difference between 45 and 29. We start with the ONES. Since 5 is less than 9, we have to borrow 1 of the TENS from the elephant. We cross off the number 5 and write 15 inside the yellow box on top of the mouse. Since we borrowed one of the Elephant's TENS, we also cross off the number 4 and subtract the borrowed 1 from 4. We write the new number 3 in the blue box on top of the elephant and subtract 9 from 15 and 2 from 3 to determine the difference of 16.

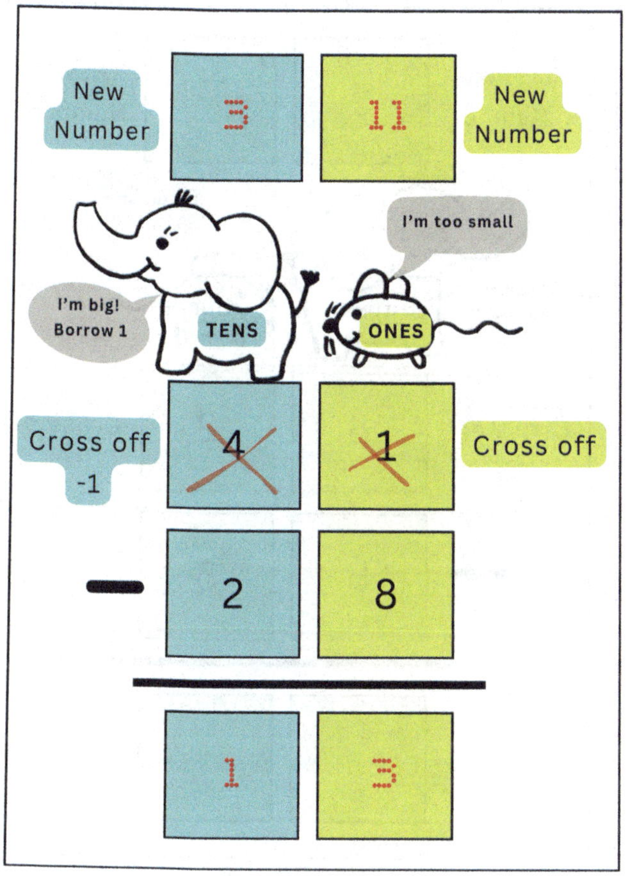
New Number
3
11
New Number
I'm too small
I'm big! Borrow 1
TENS
ONES
Cross off -1
4
1
Cross off
−
2
8
1
3

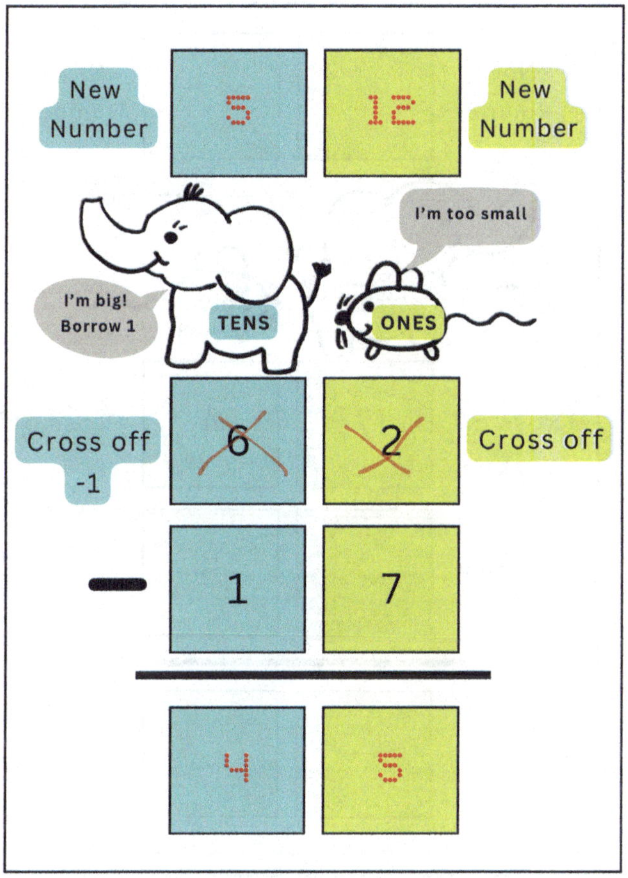
New Number
5
12
New Number
I'm too small
I'm big! Borrow 1
TENS
ONES
Cross off -1
6
2
Cross off
−
1
7
4
5

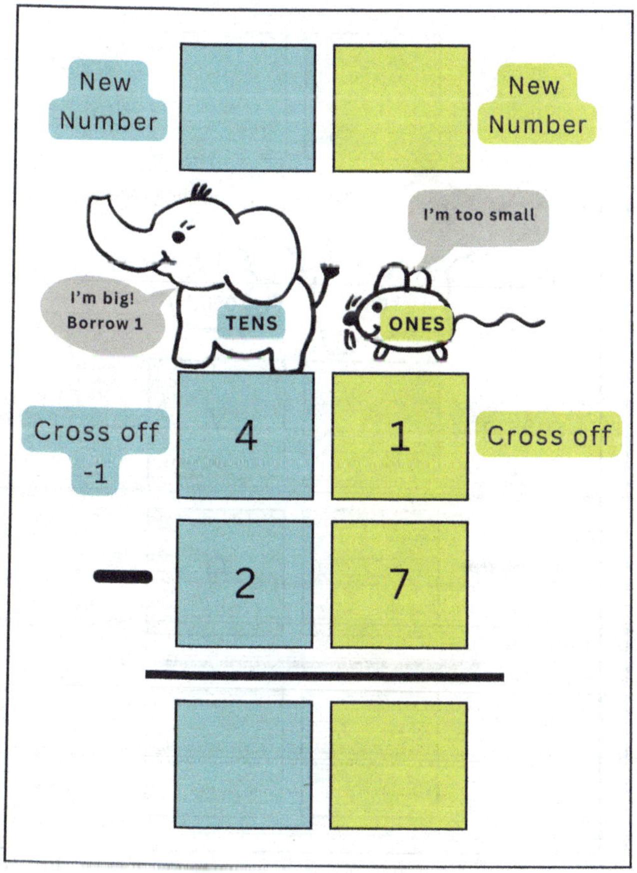
New Number
New Number
I'm too small
I'm big! Borrow 1
TENS
ONES
Cross off -1
4
1
Cross off
−
2
7

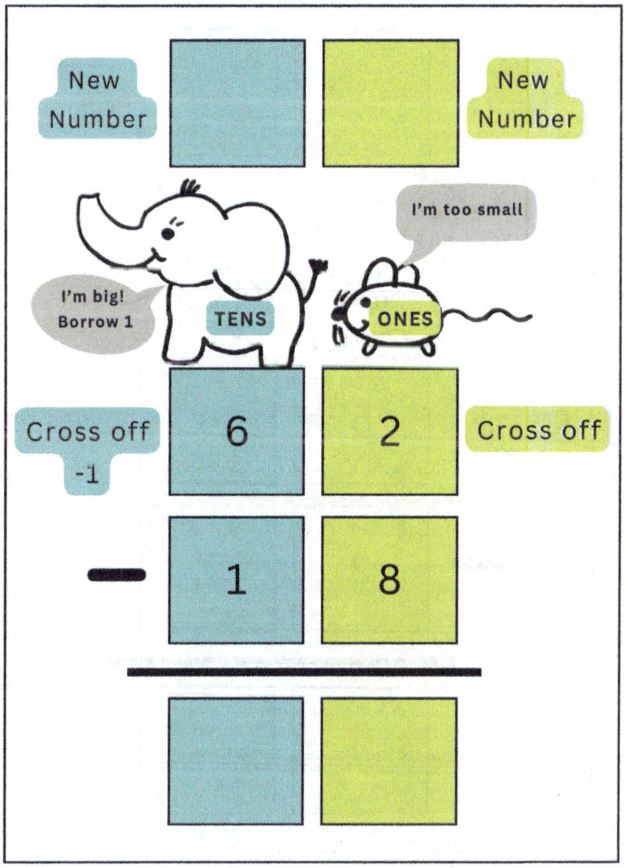
New Number
New Number
I'm too small
I'm big! Borrow 1
TENS
ONES
Cross off -1
6
2
Cross off
−
1
8

New Number | New Number

Cross off -1 | 7 | 4 | Cross off

− 2 5

New Number | New Number

Cross off -1 | 8 | 2 | Cross off

− 5 5

New Number | New Number

Cross off -1 | 5 | 4 | Cross off

− 2 7

New Number | New Number

Cross off -1 | 6 | 7 | Cross off

− 2 9

New Number | New Number

| | | | |
|---|---|---|---|
| Cross off -1 | 4 | 4 | Cross off |
| − | 1 | 8 | |
| | | | |

New Number | New Number

| | | | |
|---|---|---|---|
| Cross off -1 | 6 | 6 | Cross off |
| − | 2 | 7 | |
| | | | |

New Number | New Number

| | | | |
|---|---|---|---|
| Cross off -1 | 6 | 1 | Cross off |
| − | 2 | 6 | |
| | | | |

New Number | New Number

| | | | |
|---|---|---|---|
| Cross off -1 | 4 | 2 | Cross off |
| − | 1 | 4 | |
| | | | |

New Number | New Number

| | TENS | ONES | |
|---|---|---|---|
| Cross off -1 | 4 | 3 | Cross off |
| − | 2 | 6 | |

New Number | New Number

| | TENS | ONES | |
|---|---|---|---|
| Cross off -1 | 7 | 4 | Cross off |
| − | 3 | 7 | |

New Number | New Number

| | TENS | ONES | |
|---|---|---|---|
| Cross off -1 | 8 | 3 | Cross off |
| − | 5 | 6 | |

New Number | New Number

| | TENS | ONES | |
|---|---|---|---|
| Cross off -1 | 9 | 1 | Cross off |
| − | 2 | 4 | |

 

New Number | | | New Number

| | | | |
|---|---|---|---|
| Cross off -1 | 7 | 2 | Cross off |
| − | 4 | 6 | |
| | | | |

New Number | | | New Number

| | | | |
|---|---|---|---|
| Cross off -1 | 5 | 8 | Cross off |
| − | 3 | 9 | |
| | | | |

New Number | | | New Number

| | | | |
|---|---|---|---|
| Cross off -1 | 7 | 5 | Cross off |
| − | 4 | 6 | |
| | | | |

New Number | | | New Number

| | | | |
|---|---|---|---|
| Cross off -1 | 8 | 1 | Cross off |
| − | 3 | 7 | |
| | | | |

New Number | New Number

Cross off -1 | 6 | 7 | Cross off

− 4 8

---

New Number | New Number

Cross off -1 | 7 | 2 | Cross off

− 4 7

---

New Number | New Number

Cross off -1 | 6 | 3 | Cross off

− 4 5

---

New Number | New Number

Cross off -1 | 7 | 6 | Cross off

− 1 7

---

New Number | New Number

| Cross off -1 | 3 | 7 | Cross off |
|---|---|---|---|
| − | 1 | 8 | |

New Number | New Number

| Cross off -1 | 7 | 4 | Cross off |
|---|---|---|---|
| − | 5 | 6 | |

New Number | New Number

| Cross off -1 | 9 | 3 | Cross off |
|---|---|---|---|
| − | 7 | 5 | |

New Number | New Number

| Cross off -1 | 6 | 6 | Cross off |
|---|---|---|---|
| − | 2 | 7 | |

New Number | New Number

Cross off -1 | 5 | 5 | Cross off

− 1 8

---

New Number | New Number

Cross off -1 | 6 | 2 | Cross off

− 3 3

---

New Number | New Number

Cross off -1 | 7 | 3 | Cross off

− 5 5

---

New Number | New Number

Cross off -1 | 8 | 6 | Cross off

− 3 9

# Aquarium Column Subtraction With Regrouping

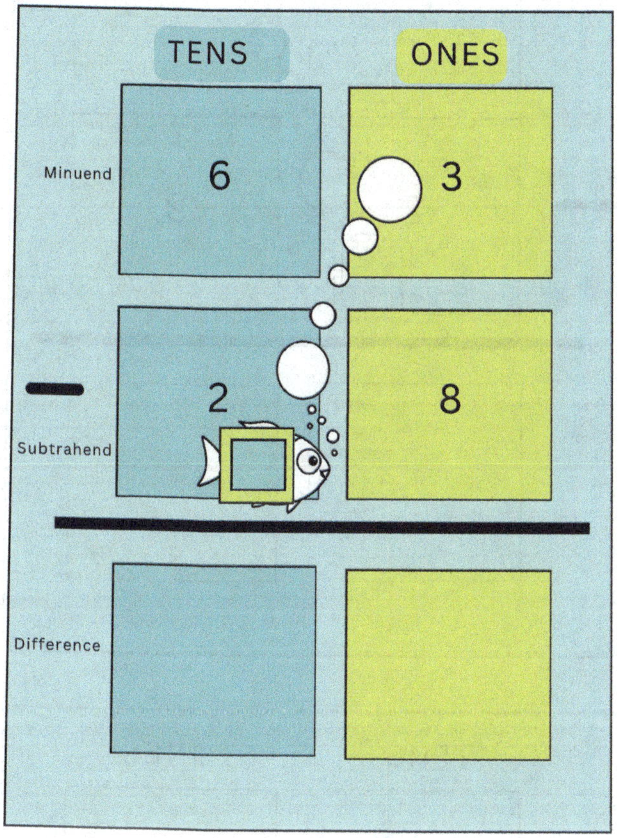


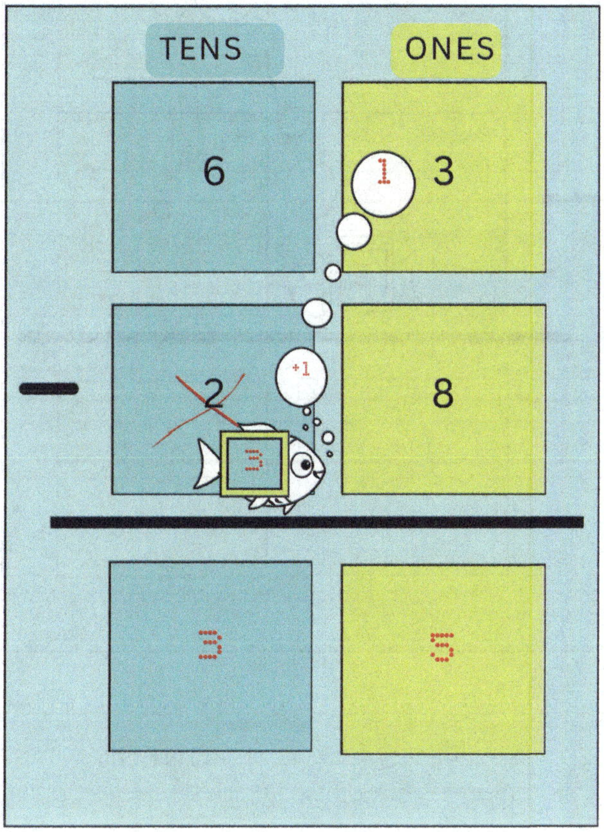


**Introduction for Parents and Teachers:**

There are various ways to explain subtraction. Find out which method works best for your students. In this method, the borrowed one is added to the TENS of the subtrahend. At the same time, your child can write the borrowed 1 in front of the ONES place of the minuend to make subtraction easier.

In our example, the students subtract 28 from 63. They start with the ONES. Since 3 is smaller than 8, they need to borrow a ten and write the borrowed 1 in the bubble in front of the ONES place, making it 13. Because 1 of the TENS was borrowed, it needs to be subtracted from the TENS, so +1 is added to the TENS place of the subtrahend. Your child can cross out the 2 and write the new number 3 (2+1) in the box in the fish.

Now they can calculate 13-8 and write the difference of 5 in the lower ONES box. Then, they move on to the TENS. Your child can now calculate 6-3 (2+1) and write the difference of 3 in the box under the TENS.

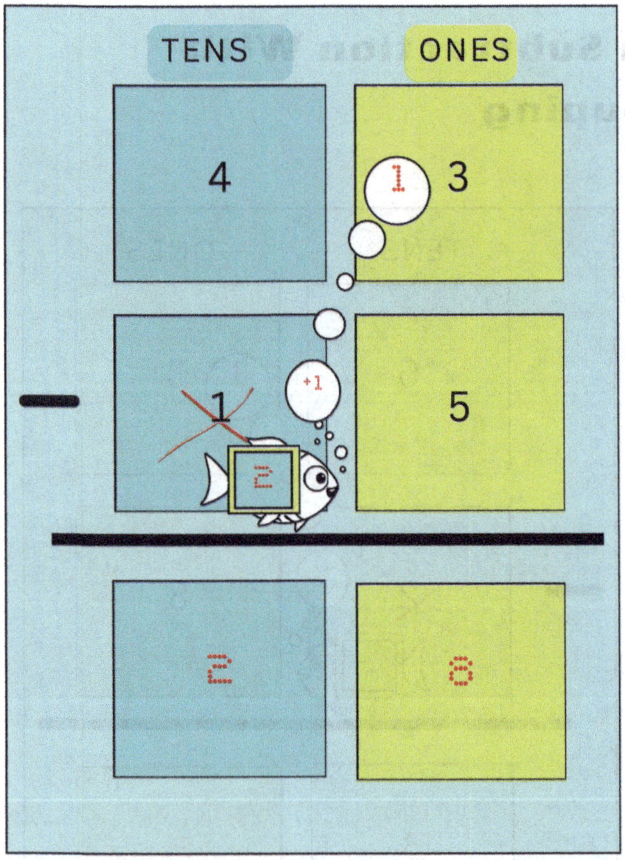
TENS
ONES
4
1
3
+1
1
2
5
2
8

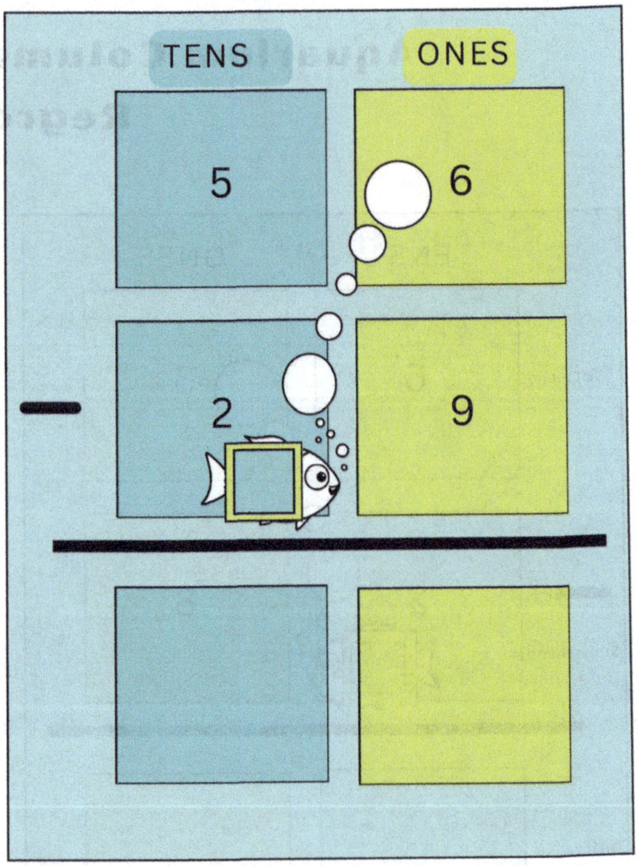
TENS
ONES
5
6
2
9

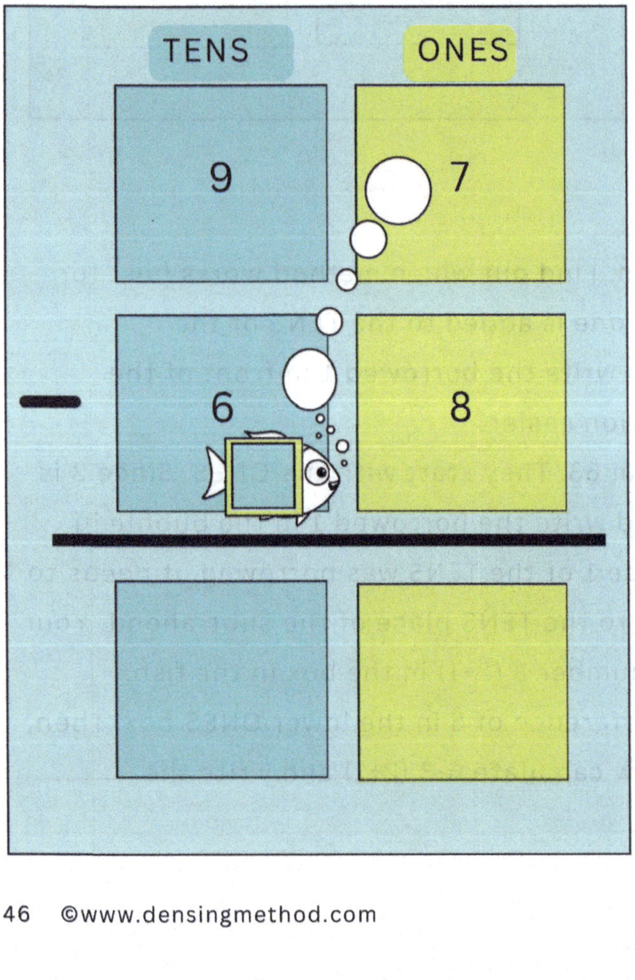
TENS
ONES
9
7
6
8

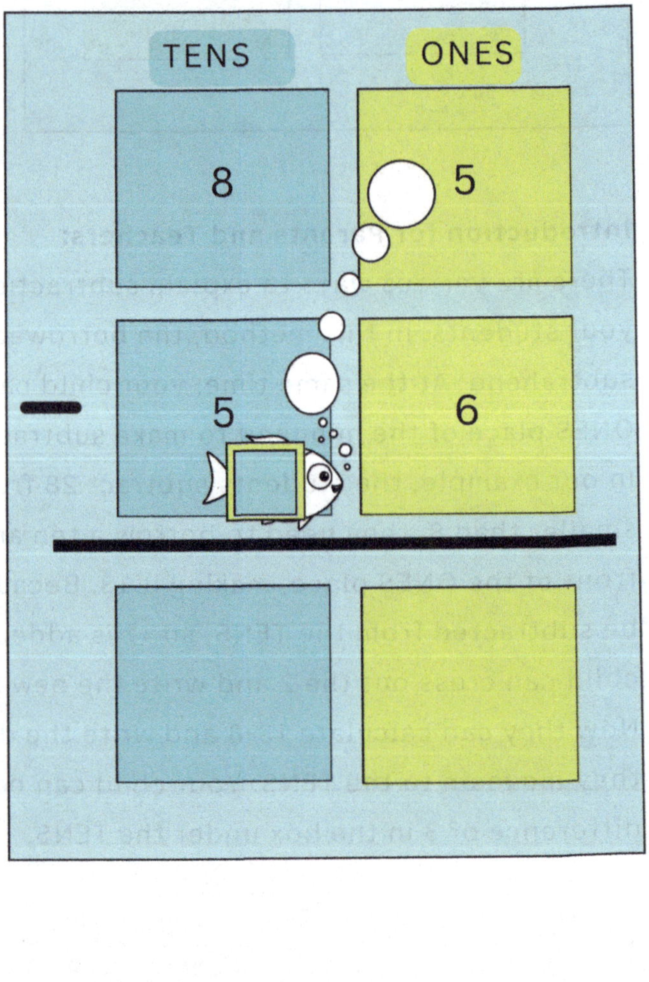
TENS
ONES
8
5
5
6

| | Tens | Ones |
|---|---|---|
| | 4 | 2 |
| − | 1 | 9 |
| | | |

| | Tens | Ones |
|---|---|---|
| | 6 | 6 |
| − | 2 | 9 |
| | | |

| | Tens | Ones |
|---|---|---|
| | 7 | 5 |
| − | 4 | 8 |
| | | |

| | Tens | Ones |
|---|---|---|
| | 5 | 2 |
| − | 3 | 6 |
| | | |

| | Tens | Ones |
|---|---|---|
| | 6 | 5 |
| − | 3 | 9 |
| | | |

| | Tens | Ones |
|---|---|---|
| | 7 | 6 |
| − | 2 | 7 |
| | | |

| | Tens | Ones |
|---|---|---|
| | 8 | 1 |
| − | 5 | 8 |
| | | |

| | Tens | Ones |
|---|---|---|
| | 4 | 3 |
| − | 2 | 6 |
| | | |

| | Tens | Ones |
|---|---|---|
| | 9 | 5 |
| − | 3 | 6 |
| | | |

| | Tens | Ones |
|---|---|---|
| | 6 | 2 |
| − | 2 | 3 |
| | | |

| | Tens | Ones |
|---|---|---|
| | 7 | 6 |
| − | 4 | 8 |
| | | |

| | Tens | Ones |
|---|---|---|
| | 8 | 3 |
| − | 4 | 4 |
| | | |

| | Tens | Ones |
|---|---|---|
| | 5 | 5 |
| − | 3 | 6 |
| | | |

| | Tens | Ones |
|---|---|---|
| | 6 | 6 |
| − | 2 | 9 |
| | | |

| | Tens | Ones |
|---|---|---|
| | 6 | 3 |
| − | 3 | 8 |
| | | |

| | Tens | Ones |
|---|---|---|
| | 3 | 3 |
| − | 1 | 9 |
| | | |

| | Tens | Ones |
|---|---|---|
| | 4 | 5 |
| − | 1 | 6 |
| | | |

| | Tens | Ones |
|---|---|---|
| | 5 | 3 |
| − | 2 | 5 |
| | | |

| | Tens | Ones |
|---|---|---|
| | 7 | 6 |
| − | 1 | 8 |
| | | |

| | Tens | Ones |
|---|---|---|
| | 6 | 3 |
| − | 3 | 4 |
| | | |

 

| | Tens | Ones |
|---|---|---|
| | 8 | 8 |
| − | 4 | 9 |
| | | |

| | Tens | Ones |
|---|---|---|
| | 5 | 6 |
| − | 2 | 9 |
| | | |

| | Tens | Ones |
|---|---|---|
| | 8 | 7 |
| − | 4 | 8 |
| | | |

| | Tens | Ones |
|---|---|---|
| | 7 | 3 |
| − | 5 | 4 |
| | | |

## Stork, Frog & Fly
## Column Addition With Regrouping Over the Numbers

**Introduction for Parents and Teachers:** When employing the traditional addition algorithm, remembering to add the carried 1 can be tricky. This worksheet addresses this challenge by providing a dedicated space over the hungry animals for the carried 1. It is color-coded and features a delightful theme of hungry animals, making it easier for children to grasp the concept. In this playful scenario, the frog eats the carried 1 from the fly, and the stork eats the carried 1 from the frog.

| | HUNDREDS | TENS | ONES |
|---|---|---|---|
| carry | 1 | 1 | |
| | 6 | 3 | 6 |
| + | 1 | 7 | 5 |
| | 8 | 1 | 1 |

| | HUNDREDS | TENS | ONES |
|---|---|---|---|
| carry | | | |
| | 4 | 7 | 5 |
| + | 3 | 3 | 9 |
| | | | |

| | HUNDREDS | TENS | ONES |
|---|---|---|---|
| carry | | | |
| | 2 | 4 | 2 |
| + | 4 | 8 | 8 |
| | | | |

| | HUNDREDS | TENS | ONES |
|---|---|---|---|
| carry | | | |
| | 5 | 7 | 5 |
| + | 2 | 4 | 7 |
| | | | |

| HUNDREDS | TENS | ONES |
|---|---|---|
| 2 | 6 | 5 |
| + 1 | 7 | 8 |
| | | |

| HUNDREDS | TENS | ONES |
|---|---|---|
| 3 | 5 | 6 |
| + 2 | 4 | 9 |
| | | |

| HUNDREDS | TENS | ONES |
|---|---|---|
| 2 | 6 | 5 |
| + 1 | 7 | 9 |
| | | |

| HUNDREDS | TENS | ONES |
|---|---|---|
| 3 | 5 | 6 |
| + 2 | 4 | 8 |
| | | |

HUNDREDS TENS ONES

| | | | |
|---|---|---|---|
| | 5 | 3 | 9 |
| + | 3 | 6 | 5 |
| | | | |

HUNDREDS TENS ONES

| | | | |
|---|---|---|---|
| | 4 | 7 | 7 |
| + | 1 | 6 | 9 |
| | | | |

HUNDREDS TENS ONES

| | | | |
|---|---|---|---|
| | 2 | 2 | 9 |
| + | 2 | 9 | 1 |
| | | | |

HUNDREDS TENS ONES

| | | | |
|---|---|---|---|
| | 4 | 7 | 7 |
| + | 3 | 7 | 8 |
| | | | |

| | HUNDREDS | TENS | ONES |
|---|---|---|---|
| | 1 | 5 | 7 |
| + | 1 | 6 | 8 |
| | | | |

| | HUNDREDS | TENS | ONES |
|---|---|---|---|
| | 1 | 6 | 6 |
| + | 2 | 4 | 7 |
| | | | |

| | HUNDREDS | TENS | ONES |
|---|---|---|---|
| | 2 | 6 | 5 |
| + | 2 | 6 | 7 |
| | | | |

| | HUNDREDS | TENS | ONES |
|---|---|---|---|
| | 3 | 7 | 6 |
| + | 3 | 3 | 5 |
| | | | |

| | HUNDREDS | TENS | ONES |
|---|---|---|---|
| | 3 | 4 | 7 |
| + | 1 | 5 | 5 |
| | | | |

| | HUNDREDS | TENS | ONES |
|---|---|---|---|
| | 4 | 7 | 5 |
| + | 2 | 4 | 8 |
| | | | |

| | HUNDREDS | TENS | ONES |
|---|---|---|---|
| | 2 | 8 | 5 |
| + | 4 | 6 | 7 |
| | | | |

| | HUNDREDS | TENS | ONES |
|---|---|---|---|
| | 3 | 6 | 4 |
| + | 4 | 5 | 8 |
| | | | |

HUNDREDS

TENS

ONES

| | | | |
|---|---|---|---|
| | 4 | 3 | 7 |
| + | 1 | 7 | 4 |
| | | | |

HUNDREDS

TENS

ONES

| | | | |
|---|---|---|---|
| | 1 | 9 | 5 |
| + | 2 | 4 | 5 |
| | | | |

HUNDREDS

TENS

ONES

| | | | |
|---|---|---|---|
| | 2 | 5 | 5 |
| + | 3 | 9 | 7 |
| | | | |

HUNDREDS

TENS

ONES

| | | | |
|---|---|---|---|
| | 3 | 6 | 5 |
| + | 4 | 6 | 8 |
| | | | |

| | HUNDREDS | TENS | ONES |
|---|---|---|---|
| | 3 | 3 | 9 |
| + | 1 | 8 | 4 |
| | | | |

| | HUNDREDS | TENS | ONES |
|---|---|---|---|
| | 5 | 8 | 5 |
| + | 2 | 4 | 7 |
| | | | |

| | HUNDREDS | TENS | ONES |
|---|---|---|---|
| | 2 | 9 | 5 |
| + | 4 | 6 | 9 |
| | | | |

| | HUNDREDS | TENS | ONES |
|---|---|---|---|
| | 4 | 6 | 3 |
| + | 3 | 5 | 8 |
| | | | |

HUNDREDS
TENS
ONES
1 3 6
+ 1 9 4

HUNDREDS
TENS
ONES
4 8 5
+ 2 3 6

HUNDREDS
TENS
ONES
4 3 2
+ 4 8 9

HUNDREDS
TENS
ONES
6 7 3
+ 2 3 7

# Stork, Frog & Fly

# Column Addition With Regrouping Under the Numbers

**Introduction for Parents and Teachers:**

There are various ways to help children remember to carry the one during addition. In this scenario, the carried-over ones are written in boxes under the numbers.

In this method, the frog "eats" the carried 1 from the fly, and the stork "eats" the carried 1 from the frog. This fun, visual approach makes it easy for children to understand and remember to add the carried numbers.

| | Hundreds | Tens | Ones |
|---|---|---|---|
| | 2 | 7 | 6 |
| + | 3 | 5 | 7 |

| | Hundreds | Tens | Ones |
|---|---|---|---|
| | 6 | 3 | 3 |

| | Hundreds | Tens | Ones |
|---|---|---|---|
| | 2 | 6 | 5 |
| + | 1 | 3 | 5 |

| | Hundreds | Tens | Ones |
|---|---|---|---|
| | | | |

| | Hundreds | Tens | Ones |
|---|---|---|---|
| | 2 | 6 | 7 |
| + | 2 | 5 | 7 |

| | Hundreds | Tens | Ones |
|---|---|---|---|
| | | | |

| | Hundreds | Tens | Ones |
|---|---|---|---|
| | 1 | 4 | 7 |
| + | 2 | 7 | 9 |

| | Hundreds | Tens | Ones |
|---|---|---|---|
| | | | |

| | Hundreds | Tens | Ones |
|---|---|---|---|
| | 5 | 7 | 5 |
| + | 2 | 7 | 7 |

| | | | |
|---|---|---|---|
| | | | |

| | Hundreds | Tens | Ones |
|---|---|---|---|
| | 2 | 6 | 7 |
| + | 4 | 6 | 5 |

| | | | |
|---|---|---|---|
| | | | |

| | Hundreds | Tens | Ones |
|---|---|---|---|
| | 2 | 8 | 9 |
| + | 4 | 9 | 7 |

| | | | |
|---|---|---|---|
| | | | |

| | Hundreds | Tens | Ones |
|---|---|---|---|
| | 3 | 4 | 8 |
| + | 2 | 9 | 9 |

| | | | |
|---|---|---|---|
| | | | |

| | Hundreds | Tens | Ones |
|---|---|---|---|
| | 3 | 7 | 5 |
| + | 2 | 3 | 6 |
| | | | |

| | Hundreds | Tens | Ones |
|---|---|---|---|
| | 7 | 7 | 5 |
| + | 1 | 3 | 7 |
| | | | |

| | Hundreds | Tens | Ones |
|---|---|---|---|
| | 5 | 6 | 8 |
| + | 2 | 6 | 7 |
| | | | |

| | Hundreds | Tens | Ones |
|---|---|---|---|
| | 4 | 8 | 7 |
| + | 2 | 7 | 7 |
| | | | |

| | Hundreds | Tens | Ones |
|---|---|---|---|
| | 2 | 2 | 5 |
| + | 2 | 8 | 6 |

| | | | |
|---|---|---|---|
| | | | |

| | Hundreds | Tens | Ones |
|---|---|---|---|
| | 1 | 6 | 7 |
| + | 4 | 9 | 3 |

| | | | |
|---|---|---|---|
| | | | |

| | Hundreds | Tens | Ones |
|---|---|---|---|
| | 4 | 8 | 3 |
| + | 4 | 3 | 7 |

| | | | |
|---|---|---|---|
| | | | |

| | Hundreds | Tens | Ones |
|---|---|---|---|
| | 5 | 4 | 2 |
| + | 2 | 6 | 9 |

| | | | |
|---|---|---|---|
| | | | |

 

| | Hundreds | Tens | Ones |
|---|---|---|---|
| | 1 | 4 | 5 |
| + | 2 | 7 | 8 |
| | | | |

| | Hundreds | Tens | Ones |
|---|---|---|---|
| | 3 | 3 | 9 |
| + | 1 | 8 | 7 |
| | | | |

| | Hundreds | Tens | Ones |
|---|---|---|---|
| | 3 | 4 | 7 |
| + | 2 | 9 | 4 |
| | | | |

| | Hundreds | Tens | Ones |
|---|---|---|---|
| | 5 | 9 | 5 |
| + | 3 | 3 | 6 |
| | | | |

# Dino Subtraction: Exploring Hundreds with Regrouping

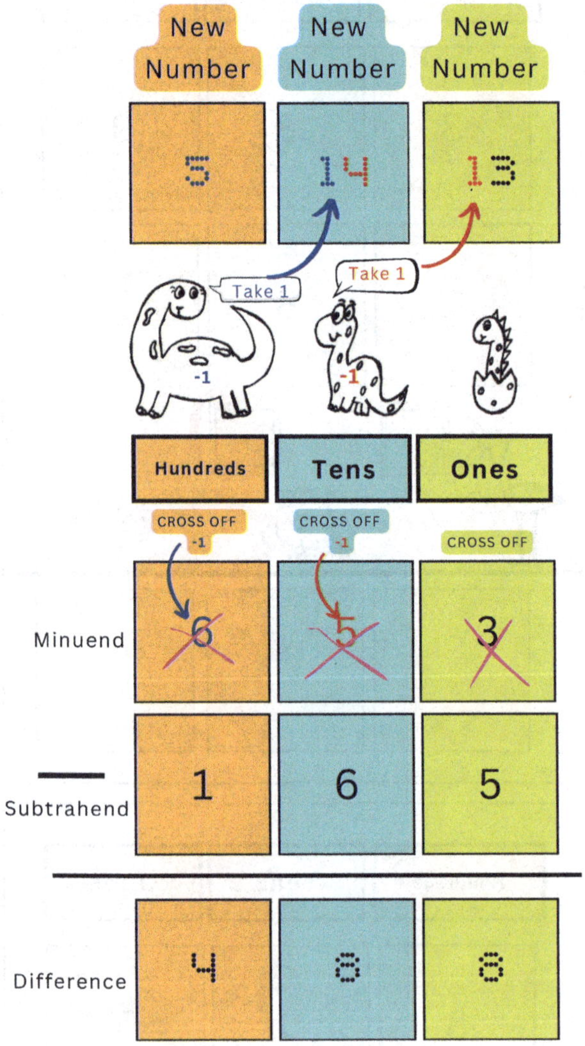


**Minuend (first number 653) - Subtrahend (second number 165) = Difference (result 488)**

These motivating dinosaur subtraction tasks, marked by size and colors, help children learn how to borrow and regroup in the hundred range. In our example, we need to determine the difference between 653 and 165. We start with the baby dinosaur in the egg, which represents the ONES. Since 3 is smaller than 5, we need to borrow one of the TENS from the larger brother dinosaur. We cross off the number 3 and write 13 in the yellow box above the baby dinosaur. Since we borrowed 1 from brother dino's TENS, we also must cross off the minuend of the TENS and subtract the borrowed 1 from 5. We write the new number 4 in the blue box above the brother dino. Now, we look at the TENS.

Since the new number 4 of the TENS is less than the number 6 of the subtrahend, we also need to take away one from the HUNDREDS.

We write this 1 in the upper box above the brother dino and now get the number 14. Since we took 1 from Mama Dino, we cross off number 6 and subtract 1. We write our new number 5 in the orange box above the dino mom. Now we can subtract the numbers. We start with the ONES and subtract 5 from 13 and get 8. Then we look at the TENS and subtract 6 from 14 and get 8. Finally, we look at the HUNDREDS and subtract 1 from 5 and get 4. Our result is 488.

Subtraction tends to be more challenging for children to comprehend and master. However, as they engage in repeated practice, the concept becomes clearer over time. There are diverse teaching methods for subtraction. This approach provides visual aids, which can aid children in understanding the concept.

# Dino Subtraction: Exploring Hundreds with Regrouping

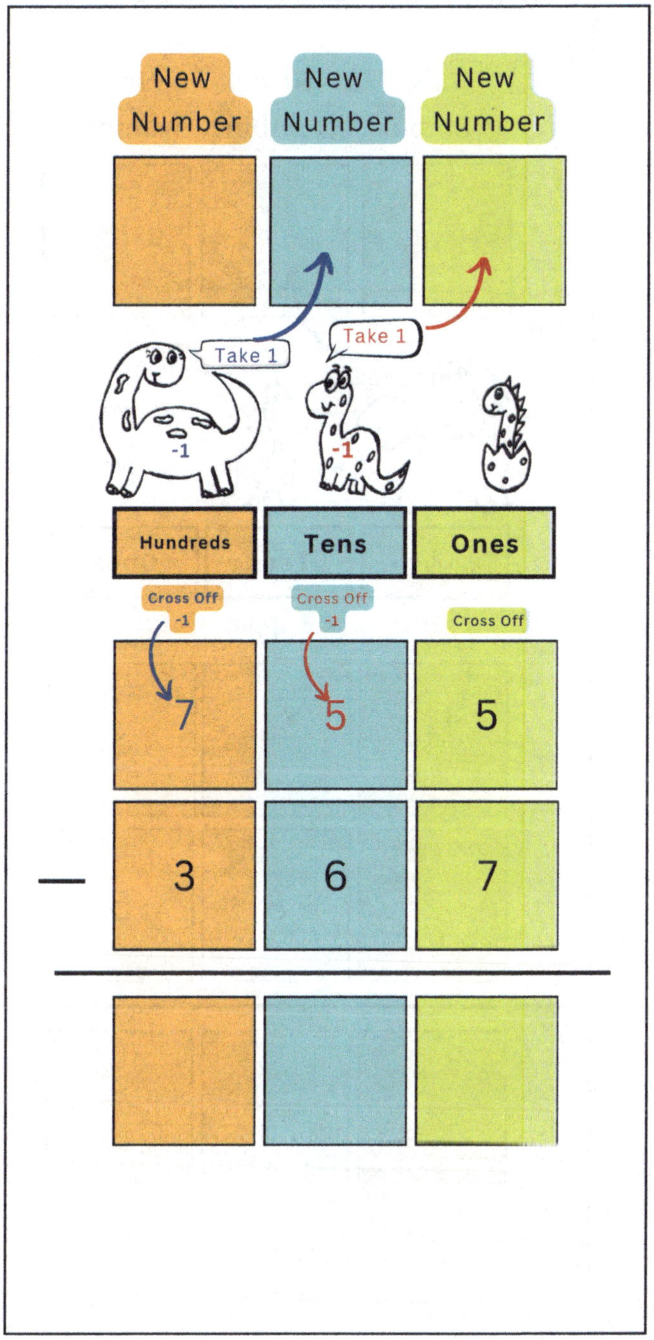

# Dino Subtraction: Exploring Hundreds with Regrouping

| New Number | New Number | New Number |
|---|---|---|
| | | |

| Hundreds | Tens | Ones |
|---|---|---|
| Cross Off -1 | Cross Off -1 | Cross Off |
| 4 | 2 | 3 |
| − 2 | 8 | 5 |
| | | |

| New Number | New Number | New Number |
|---|---|---|
| | | |

| Hundreds | Tens | Ones |
|---|---|---|
| Cross Off -1 | Cross Off -1 | Cross Off |
| 7 | 6 | 5 |
| − 4 | 7 | 8 |
| | | |

 

# Dino Subtraction: Exploring Hundreds with Regrouping

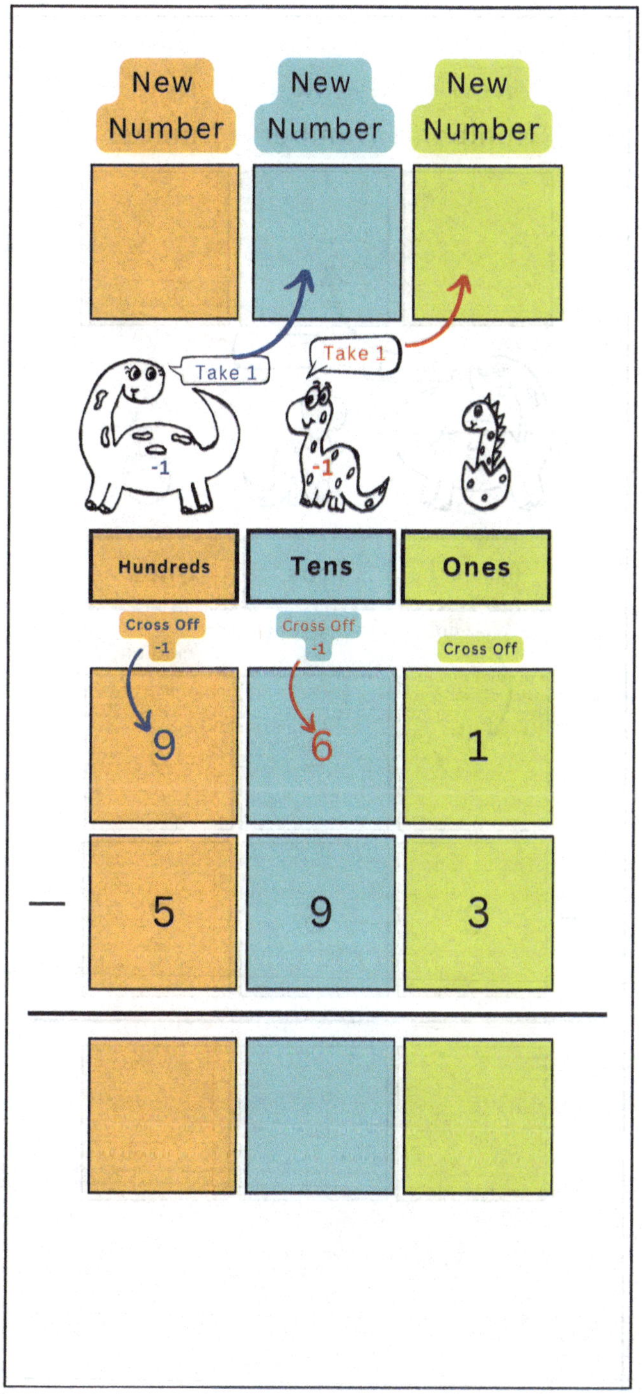

# Dino Subtraction: Exploring Hundreds with Regrouping

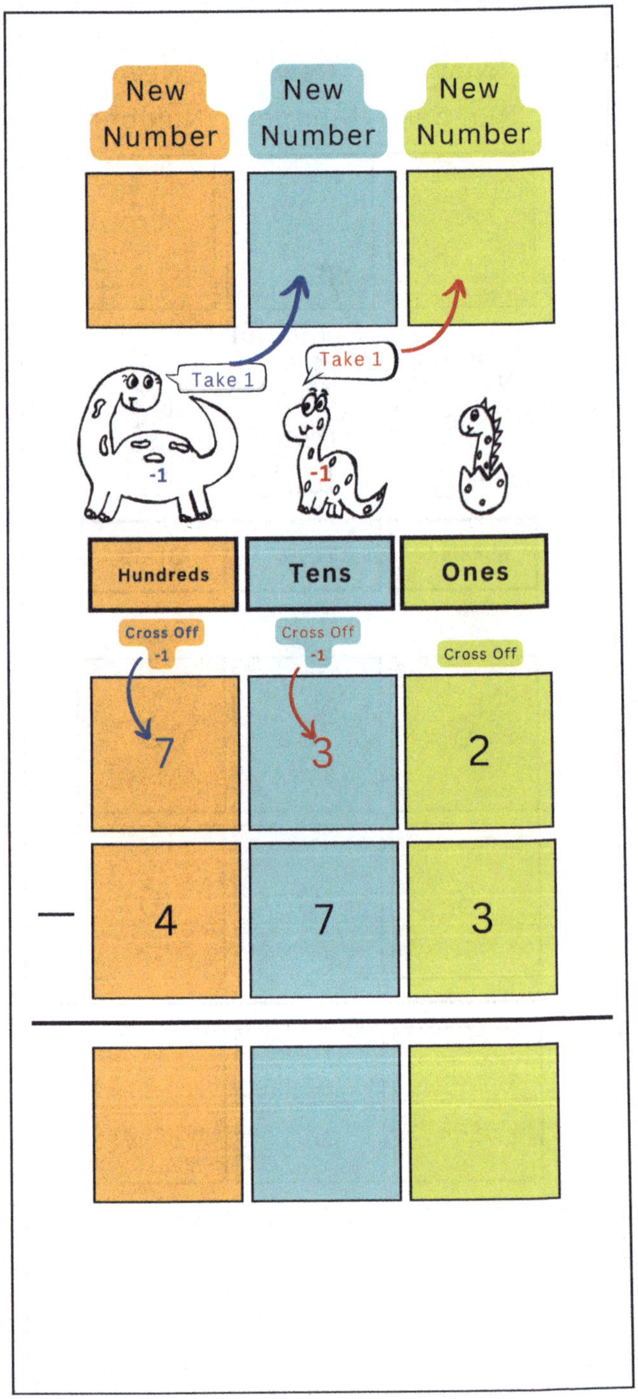

# Dino Subtraction: Exploring Hundreds with Regrouping

# Dino Subtraction: Exploring Hundreds with Regrouping

# Dino Subtraction: Exploring Hundreds with Regrouping

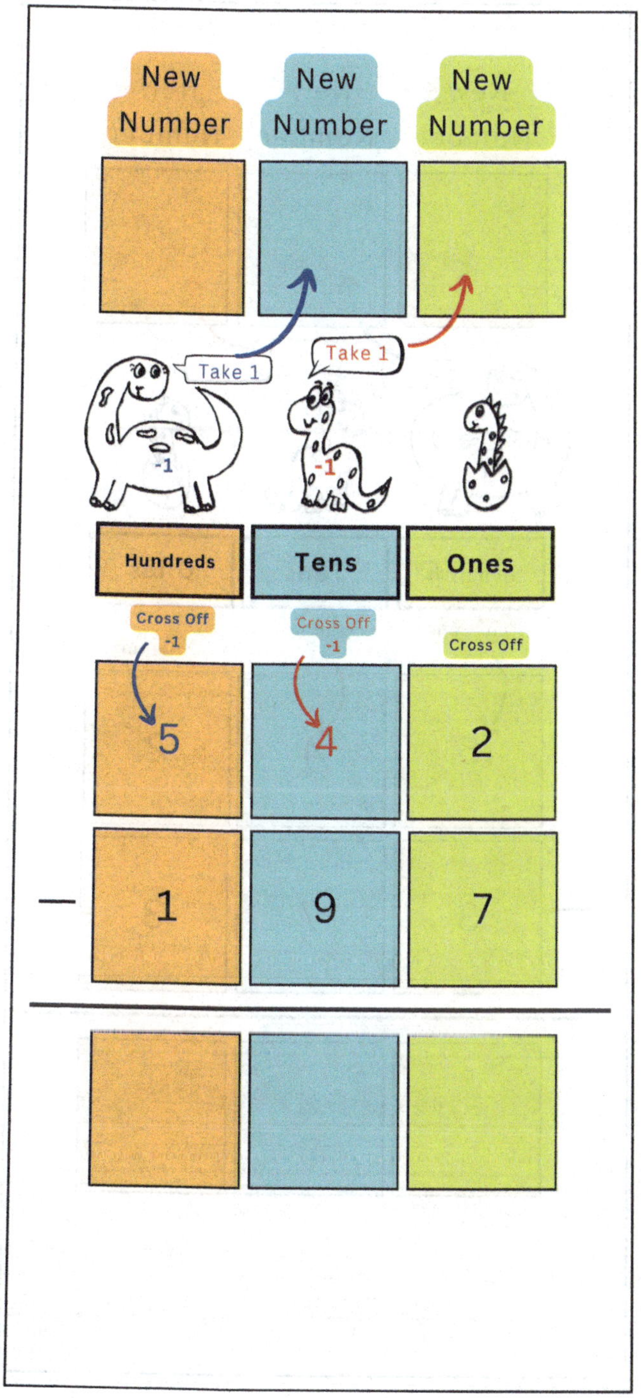

# Dino Subtraction: Exploring Hundreds with Regrouping

| New Number | New Number | New Number |
|---|---|---|
| | | |

| | Hundreds | Tens | Ones |
|---|---|---|---|
| | Cross Off -1 | Cross Off -1 | Cross Off |
| | 7 | 5 | 2 |
| − | 3 | 8 | 5 |
| | | | |

| New Number | New Number | New Number |
|---|---|---|
| | | |

| | Hundreds | Tens | Ones |
|---|---|---|---|
| | Cross Off -1 | Cross Off -1 | Cross Off |
| | 5 | 4 | 2 |
| − | 3 | 7 | 3 |
| | | | |

# Dino Subtraction: Exploring Hundreds with Regrouping

# Dino Subtraction: Exploring Hundreds with Regrouping

| New Number | New Number | New Number |
|---|---|---|
| | | |

| Hundreds | Tens | Ones |
|---|---|---|
| Cross Off -1 | Cross Off -1 | Cross Off |
| 5 | 6 | 2 |
| − 1 | 7 | 5 |
| | | |

| New Number | New Number | New Number |
|---|---|---|
| | | |

| Hundreds | Tens | Ones |
|---|---|---|
| Cross Off -1 | Cross Off -1 | Cross Off |
| 8 | 2 | 5 |
| − 4 | 3 | 7 |
| | | |

 

# Dino Subtraction: Exploring Hundreds with Regrouping

# Dino Subtraction: Exploring Hundreds with Regrouping

# Worm Subtraction: Exploring Hundreds with Regrouping

**Introduction for Parents and Teachers:**

**Minuend (top number 767) - Subtrahend (bottom number 388) = Difference (result 379)**

In this subtraction method, the borrowed 1 is added to the TENS and HUNDREDS places of the subtrahend. At the same time, students can write the borrowed 1 in front of the ONES and TENS places of the minuend to make subtraction easier. For example, when subtracting 388 from 767, remind your child starts with the ONES. Since 7 is smaller than 8, they borrow one of the TENS, writing the borrowed 1 in the yellow oval in front of the ONES, making it 17.

Because 1 of the TENS was borrowed, it needs to be subtracted from the TENS, so +1 is added to the TENS place of the subtrahend. The child crosses out the 8 and writes 9 (8+1) in the blue lower oval.

Looking at the TENS, 6 is smaller than the 9. The child has to borrow from the HUNDREDS, writing the borrowed 1 in the blue oval next to the 6, making it 16. Then, they add +1 to the HUNDREDS place of the subtrahend, changing 3 to 4 in the orange lower oval.

Now, they can subtract step by step:

- Start with the ONES: Calculate 17 - 8 and get 9.
- Move to the TENS: Calculate 16 - 9 and get 7.
- End with the HUNDREDS: Calculate 7 - 4 to get 3. The result is 379.

This method helps children break down the subtraction process into manageable steps. Subtraction can take time and practice for a child to master. It's important to repeat the method that makes the most sense to your child to help them understand and remember it better.

Hundreds
Tens
Ones
8
1
6
1
5
—
3
+1
7
+1
8
4
8
4
8
7

Hundreds
Tens
Ones
7
4
7
—
2
6
9

Hundreds
Tens
Ones
9
1
6
—
5
5
8

Hundreds
Tens
Ones
7
6
7
—
3
8
8

| | Hundreds | Tens | Ones |
|---|---|---|---|
| | 7 | 2 | 2 |
| − | 4 | 4 | 4 |
| | | | |

| | Hundreds | Tens | Ones |
|---|---|---|---|
| | 6 | 7 | 2 |
| − | 2 | 8 | 6 |
| | | | |

| | Hundreds | Tens | Ones |
|---|---|---|---|
| | 8 | 4 | 6 |
| − | 2 | 6 | 8 |
| | | | |

| | Hundreds | Tens | Ones |
|---|---|---|---|
| | 4 | 4 | 7 |
| − | 1 | 7 | 8 |
| | | | |

| | Hundreds | Tens | Ones |
|---|---|---|---|
| | 5 | 4 | 1 |
| − | 3 | 5 | 4 |
| | | | |

| | Hundreds | Tens | Ones |
|---|---|---|---|
| | 8 | 1 | 2 |
| − | 3 | 5 | 6 |
| | | | |

| | Hundreds | Tens | Ones |
|---|---|---|---|
| | 8 | 3 | 4 |
| − | 2 | 5 | 7 |
| | | | |

| | Hundreds | Tens | Ones |
|---|---|---|---|
| | 6 | 6 | 2 |
| − | 3 | 7 | 8 |
| | | | |

 

| | Hundreds | Tens | Ones |
|---|---|---|---|
| | 5 | 3 | 2 |
| − | 2 | 4 | 7 |
| | | | |

| | Hundreds | Tens | Ones |
|---|---|---|---|
| | 4 | 5 | 1 |
| − | 1 | 8 | 7 |
| | | | |

| | Hundreds | Tens | Ones |
|---|---|---|---|
| | 7 | 2 | 5 |
| − | 3 | 7 | 8 |
| | | | |

| | Hundreds | Tens | Ones |
|---|---|---|---|
| | 9 | 1 | 2 |
| − | 1 | 4 | 8 |
| | | | |

| | Hundreds | Tens | Ones |
|---|---|---|---|
| | 3 | 2 | 3 |
| − | 1 | 5 | 4 |
| | | | |

| | Hundreds | Tens | Ones |
|---|---|---|---|
| | 6 | 4 | 2 |
| − | 4 | 5 | 3 |
| | | | |

| | Hundreds | Tens | Ones |
|---|---|---|---|
| | 9 | 4 | 6 |
| − | 6 | 5 | 8 |
| | | | |

| | Hundreds | Tens | Ones |
|---|---|---|---|
| | 5 | 5 | 5 |
| − | 3 | 7 | 6 |
| | | | |

| | Hundreds | Tens | Ones |
|---|---|---|---|
| | 5 | 2 | 6 |
| − | 3 | 4 | 9 |
| | | | |

| | Hundreds | Tens | Ones |
|---|---|---|---|
| | 9 | 7 | 2 |
| − | 2 | 8 | 3 |
| | | | |

| | Hundreds | Tens | Ones |
|---|---|---|---|
| | 5 | 4 | 3 |
| − | 2 | 5 | 8 |
| | | | |

| | Hundreds | Tens | Ones |
|---|---|---|---|
| | 3 | 6 | 5 |
| − | 1 | 7 | 6 |
| | | | |

| | Hundreds | Tens | Ones |
|---|---|---|---|
| | 3 | 2 | 3 |
| − | 1 | 5 | 4 |
| | | | |

| | Hundreds | Tens | Ones |
|---|---|---|---|
| | 6 | 4 | 2 |
| − | 4 | 5 | 3 |
| | | | |

| | Hundreds | Tens | Ones |
|---|---|---|---|
| | 9 | 4 | 6 |
| − | 6 | 5 | 8 |
| | | | |

| | Hundreds | Tens | Ones |
|---|---|---|---|
| | 5 | 5 | 5 |
| − | 3 | 7 | 6 |
| | | | |

 

# Books by the Author H. Densing

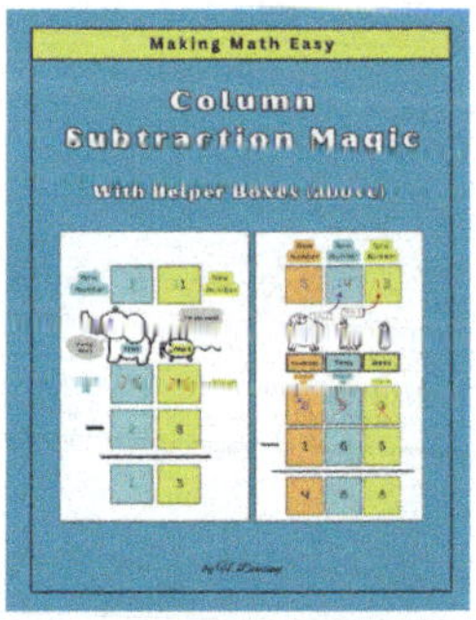

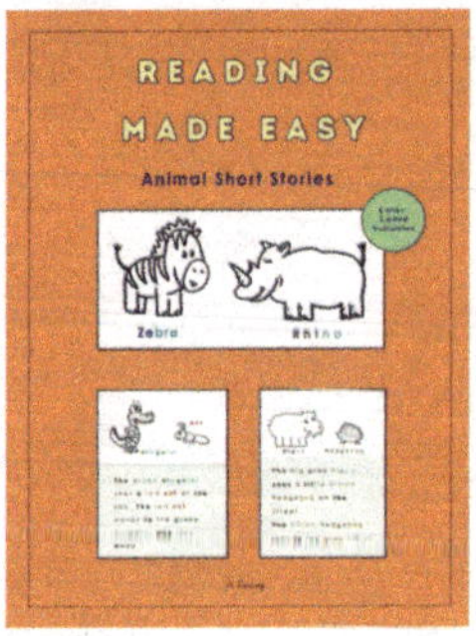

Made in the USA
Monee, IL
02 November 2026

67386638R00050